ENERGY SECTOR STANDARD OF THE PEOPLE' S REPUBLIC OF CHINA

中华人民共和国能源行业标准

Specification for Preparation of Safety Pre-assessment Report for Hydropower Projects

水电工程安全预评价报告编制规程

NB/T 35015-2021

Replace NB/T 35015-2013

Chief Development Department: China Renewable Energy Engineering Institute

Approval Department: National Energy Administration of the People's Republic of China

Implementation Date: July 1, 2021

U0291727

China Water & Power Press

中国水利水电出版社

Beijing 2024

All rights reserved. No part of this publication may be reproduced, stored in a retrieval system, or transmitted in any form or by any means—electronic, mechanical, photocopying, recording or otherwise, without prior written permission of the publisher.

图书在版编目（CIP）数据

水电工程安全预评价报告编制规程 : NB/T 35015-2021 = Specification for Preparation of Safety Pre-assessment Report for Hydropower Projects（NB/T 35015-2021） : 英文 / 国家能源局发布. -- 北京 : 中国水利水电出版社, 2024. 10. -- ISBN 978-7-5226-2814-1

Ⅰ. TV513-65

中国国家版本馆CIP数据核字第2024KV0841号

ENERGY SECTOR STANDARD
OF THE PEOPLE'S REPUBLIC OF CHINA
中华人民共和国能源行业标准

Specification for Preparation of Safety Pre-assessment Report for Hydropower Projects

水电工程安全预评价报告编制规程

NB/T 35015-2021

Replace NB/T 35015-2013

（英文版）

Issued by National Energy Administration of the People's Republic of China
国家能源局　发布
Translation organized by China Renewable Energy Engineering Institute
水电水利规划设计总院　组织翻译
Published by China Water & Power Press
中国水利水电出版社　出版发行
Tel: (+ 86 10) 68545888　68545874
sales@mwr.gov.cn
Account name: China Water & Power Press
Address: No.1, Yuyuantan Nanlu, Haidian District, Beijing 100038, China
http: //www.waterpub.com.cn
中国水利水电出版社微机排版中心　排版
北京中献拓方科技发展有限公司　印刷
184mm×260mm　16 开本　2 印张　63 千字
2024 年 10 月第 1 版　2024 年 10 月第 1 次印刷
Price（定价）: ¥ 330.00

Introduction

This English version is one of China's energy sector standard series in English. Its translation was organized by China Renewable Energy Engineering Institute authorized by National Energy Administration of the People's Republic of China in compliance with relevant procedures and stipulations. This English version was issued by National Energy Administration of the People's Republic of China in Announcement [2023] No. 8 dated December 28, 2023.

This version was translated from the Chinese Standard NB/T 35015-2021, *Specification for Preparation of Safety Pre-assessment Report for Hydropower Projects,* published by China Water & Power Press. The copyright is reserved by National Energy Administration of the People's Republic China. In the event of any discrepancy in the implementation, the Chinese version shall prevail.

Many thanks go to the staff from the relevant standard development organizations and those who have provided generous assistance in the translation and review process.

For further improvement of the English version, any comments and suggestions are welcome and should be addressed to:

China Renewable Energy Engineering Institute
No. 2 Beixiaojie, Liupukang, Xicheng District, Beijing 100120, China
Website: www.creei.cn

Translating organizations:

POWERCHINA Northwest Engineering Corporation Limited

China Renewable Energy Engineering Institute

Translating staff:

LI Mao   LU Huayan   SUN Yilin   LI Zhongjie

JIA Chao

Review panel members:

LIU Xiaofen	POWERCHINA Zhongnan Engineering Corporation Limited
GUO Jie	POWERCHINA Beijing Engineering Corporation Limited
YE Bin	POWERCHINA Huadong Engineering Corporation Limited

QIE Chunsheng	Senior English Translator
YAN Wenjun	Army Academy of Armored Forces, PLA
LI Hong	POWERCHINA Northwest Engineering Corporation Limited

National Energy Administration of the People's Republic of China

翻译出版说明

本译本为国家能源局委托水电水利规划设计总院按照有关程序和规定，统一组织翻译的能源行业标准英文版系列译本之一。2023 年 12 月 28 日，国家能源局以 2023 年第 8 号公告予以公布。

本译本是根据中国水利水电出版社出版的《水电工程安全预评价报告编制规程》NB/T 35015—2021 翻译的，著作权归国家能源局所有。在使用过程中，如出现异议，以中文版为准。

本译本在翻译和审核过程中，本标准编制单位及编制组有关成员给予了积极协助。

为不断提高本译本的质量，欢迎使用者提出意见和建议，并反馈给水电水利规划设计总院。

地址：北京市西城区六铺炕北小街 2 号
邮编：100120
网址：www.creei.cn

本译本翻译单位：中国电建集团西北勘测设计研究院有限公司
水电水利规划设计总院

本译本翻译人员：李　茂　卢花妍　孙艺林　李仲杰
贾　超

本译本审核人员：

刘小芬　中国电建集团中南勘测设计研究院有限公司

郭　洁　中国电建集团北京勘测设计研究院有限公司

叶　彬　中国电建集团华东勘测设计研究院有限公司

郄春生　英语高级翻译

闫文军　中国人民解放军陆军装甲兵学院

李　宏　中国电建集团西北勘测设计研究院有限公司

国家能源局

Announcement of National Energy Administration of the People's Republic of China [2021] No. 1

National Energy Administration of the People's Republic of China has approved and issued 320 energy sector standards including *Code for Integrated Resettlement Design of Hydropower Projects* (Attachment 1), the foreign language versions of 113 energy sector standards including *Carbon Steel and Low Alloy Steel for Pressurized Water Reactor Nuclear Power Plants—Part 7: Class 1, 2, 3 Plates* (Attachment 2), and the amendment notification for 5 energy sector standards including *Technical Code for Investigation and Assessment of Aquatic Ecosystem for Hydropower Projects* (Attachment 3).

Attachments: 1. Directory of Sector Standards

2. Directory of Foreign Language Versions of Sector Standards

3. Amendment Notification for Sector Standards

National Energy Administration of the People's Republic of China

January 7, 2021

Attachment 1:

Directory of Sector Standards

Serial number	Standard No.	Title	Replaced standard No.	Adopted international standard No.	Approval date	Implementation date
…						
206	NB/T 35015-2021	Specification for Preparation of Safety Pre-assessment Report for Hydropower Projects	NB/T 35015-2013		2021-01-07	2021-07-01
…						

Foreword

According to the requirements of the *Notice on Releasing the Amendment and Adjustment Plan for the Development and Revision of Energy Sector Standards in 2019 issued by the General Affairs Department of National Energy Administration of the People's Republic of China*, and after extensive investigation and research, summarization of practical experience, and wide solicitation of opinions, the drafting group has prepared this specification.

The main technical contents of this specification include: general provisions, terms, basic requirements, requirements for preparation of report basic contents.

The main technical contents revised are as follows:

—Deleting the filing requirements for the safety pre-assessment report of hydropower projects;

—Deleting the "photocopy of the safety assessment qualification certificate" and "safety assessment qualification certificate number" in the original "Appendix C Format of Safety Pre-assessment Report".

National Energy Administration of the People's Republic of China is in charge of the administration of this specification. China Renewable Energy Engineering Institute has proposed this specification and is responsible for its routine management. Energy Sector Standardization Technical Committee on Hydropower Investigation and Design (NEA/TC15) is responsible for the explanation of specific technical contents. Comments and suggestions in the implementation of this specification should be addressed to:

China Renewable Energy Engineering Institute
No.2 Beixiaojie, Liupukang, Xicheng District, Beijing 100120, China

Chief development organizations:

POWERCHINA Northwest Engineering Corporation Limited

China Renewable Energy Engineering Institute

Participating development organizations:

China Water Resources & Hydropower Engineering Consulting Corporation

POWERCHINA Huadong Engineering Corporation Limited

POWERCHINA Zhongnan Engineering Corporation Limited

POWERCHINA Kunming Engineering Corporation Limited

Beijing Millennium Engineering Technology Co., Ltd.

Chief drafting staff:

LI Jing, YANG Zhigang, LI Mao, LIU Yunfeng

WU Xi, LI Yuqin, YAO Yunlong, LU Huayan

LU Wenwen, LI Weiwei, NIU Wenbin, WANG Jilin

JIA Chao, ZHENG Xingang, ZHENG Lin, GUO Chen

FENG Zhenqiu, JIANG Hanren, ZHU Zhe, ZHANG Xiaoli

PAN Jian, YUAN Dongcheng, DING Bo, WU Dexin

Review panel members:

HE Haiyuan, WU Maolin, TIAN Dongsheng, CAO Yiming

LI Deyu, LIU Shihuang, CHEN Yinqi, PAN Shufa

JIA Juntian, ZHU Jun, ZENG Hui, ZHANG Xiaoguang

ZHANG Yan, LIU Rongli, LI Shisheng

Contents

1 General Provisions

1.0.1 This specification specifies the content, format, and level of detail for the preparation of safety pre-assessment reports for hydropower projects.

1.0.2 This specification is applicable to the preparation of safety pre-assessment reports for hydropower projects.

1.0.3 In addition to this specification, the preparation of safety pre-assessment reports for hydropower projects shall comply with other current relevant standards of China.

2 Terms

2.0.1 accident

unexpected event causing death, disease, damage, injury, or other losses

2.0.2 hazardous and harmful factors

factors which might cause casualties, health problems even diseases to people, or damage to equipment and facilities

2.0.3 production process

entire process in which workers are engaged in production activities in workplaces, including safe operation and construction

2.0.4 workplace

place and space where workers are engaged in occupational activities

2.0.5 emergency response plan

emergency preparedness plan to minimize damage in the event of an accident

2.0.6 assessment unit

individual part for safety assessment, which is a component of production process or place divided according to characteristics

2.0.7 major hazard installation

unit for long-term or temporary production, transport, use, or storage of hazardous goods, whose amount is equal to or exceeds the critical amount

2.0.8 hazard

dangerous state of objects, unsafe human behaviors or management deficiencies that might lead to accidents

3 Basic Requirements

3.0.1 The safety pre-assessment report of a hydropower project shall be prepared in the feasibility study stage of the project.

3.0.2 The safety pre-assessment procedure of a hydropower project shall include preparations, identification and analysis of hazardous and harmful factors, division of assessment into units, qualitative and quantitative assessment, determination of risk level, provision of suggestions on safety risk control measures, conclusion of safety assessment, and preparation of the safety pre-assessment report.

3.0.3 The safety pre-assessment report of a hydropower project shall be prepared on the basis of on-site survey by safety assessment personnel, collection and analysis of the related data and accident cases of hydropower projects, and engineering analogy. The on-site survey shall mainly cover the natural conditions, geological conditions and surrounding social environment of the project site area. The data list for safety pre-assessment of hydropower project should comply with Appendix A of this specification.

3.0.4 The contents for preparation of safety pre-assessment report for a hydropower project shall comply with Appendix B of this specification.

3.0.5 The format for preparation of safety pre-assessment report for a hydropower project shall comply with Appendix C of this specification.

4 Requirements for Preparation of Report Basic Contents

4.1 Introduction

4.1.1 The introduction in the safety pre-assessment report of a hydropower project shall cover the assessment purpose, scope, working process and basis, and a profile of the project owner.

4.1.2 The purpose, scope and work process of safety pre-assessment of a hydropower project shall meet the following requirements:

1 The safety pre-assessment shall identify and analyze the potential hazardous and harmful factors of the hydropower project according to the relevant basic data; predict the possibility and severity of accidents and determine risk levels in accordance with the requirements of work safety laws, regulations and national or sector standards; put forward scientific, reasonable and feasible safety risk control measures; and make conclusions of safety pre-assessment.

2 The safety pre-assessment shall cover the whole process from design, construction to operation, maintenance and management of the hydropower project. When a renovation or extension project shares parts with existing facilities, the scope of safety pre-assessment shall also include the shared parts.

3 The safety pre-assessment report shall describe the preparations and working process, and provide the block diagram of assessment procedure.

4.1.3 The basis for preparing the safety pre-assessment report of a hydropower project shall include:

1 National laws and regulations, and national or sector standards.

2 Pre-feasibility study report and its review opinion.

3 Interim results of feasibility study report and relevant data.

4.1.4 The introduction in the safety pre-assessment report of a hydropower project shall briefly describe the basic information and business scope of the project owner.

4.2 Project Overview

4.2.1 The project overview in the safety pre-assessment report shall cover the following project information: project location, surroundings, hydrometeorology, sediment, engineering geology, project tasks and scale, project layout and buildings (structures), electromechanical equipment and hydraulic steel structures, fire protection, construction, project investment, project characteristics table, etc.

4.2.2 The overview of renovation or extension works of a hydropower station shall also cover the original production scale, production process, general layout, access roads, and the utilization of existing site, buildings(structures), equipment and facilities.

4.3 Identification and Analysis of Hazardous and Harmful Factors and Major Hazard Installations

4.3.1 The safety pre-assessment report of a hydropower project shall list the basis for identification and analysis of hazardous and harmful factors.

4.3.2 The safety pre-assessment report of a hydropower project shall, within the scope of assessment or hydropower complex boundary, identify the potential hazardous and harmful factors and their locations and damage modes in accordance with the current national standards GB/T 13861, *Classification and Code for the Hazardous and Harmful Factors in Process* and GB 6441, *The Classification for Casualty Accidents of Enterprise Staff and Workers* in terms of surrounding conditions, dam siting, general layout, roads and transportation, buildings (structures), electromechanical equipment, working environment, safety management, natural disasters, construction process, engineering analogy, and the accumulated actual data of existing projects and typical accident cases. The identification and analysis of hazardous and harmful factors and major hazard installations shall also meet the following requirements:

1 The hazardous and harmful factors in the production process of the hydropower project shall be identified and analyzed comprehensively and accurately.

2 The safety pre-assessment report shall also identify and analyze the hazardous and harmful factors in the construction period.

3 The typical accident cases shall be highly pertinent.

4.3.3 The safety pre-assessment report of a hydropower project shall identify and analyze the hazardous goods involved in the production process in accordance with the current national standard GB 18218, *Identification of Major Hazard Installations for Hazardous Chemicals*, and determine the major hazard installations and their hazard degree in the construction and operation of the project.

4.4 Assessment Units and Assessment Methods

4.4.1 The safety pre-assessment report shall explain the principles for defining assessment units.

4.4.2 The safety pre-assessment units should be classified into the project siting and general layout unit, near-dam reservoir bank unit, water retaining structure unit, navigation structure unit, water release system unit, headrace system unit, powerhouse and power generation system unit, tailrace system unit, switchyard unit, access roads unit, log pass unit, fish pass unit, safety monitoring system unit, special equipment unit, working environment unit, construction unit, and safety management unit. The assessment units may also be defined according to individual works or the category of hazardous and harmful factors. Assessment units shall cover all the assessment content and hazardous and harmful factors within the assessment scope.

4.4.3 Appropriate safety assessment methods shall be selected according to the assessment purpose, requirements, and project characteristics; a brief description shall be made on the selected assessment methods and the reason to select the methods.

4.4.4 The names and locations of potential hazardous and harmful factors of each assessment unit and the assessment methods shall be summarized and described in the safety pre-assessment of the hydropower project.

4.5 Qualitative and Quantitative Assessment

4.5.1 The safety pre-assessment shall employ the defined assessment method to evaluate the possibility and severity of accidents caused by potential hazards and harmful factors existing in each unit, and determine the risk level. The risk levels shall be classified, from high to low, into the major risk, high risk, moderate risk, and low risk.

4.5.2 Quantitative safety assessment shall be carried out for the following items.

1 Instability of dam or main structures;

2 Fire simulation calculation for underground caverns.

4.5.3 The hydropower project safety pre-assessment shall include the project siting and general layout, earthquake, reservoir-induced earthquake, debris flow, slope failure, reservoir leakage, instability of surrounding rock, floods, sediment deposition, water retaining structures, navigation structures, headrace structures, water release structures, powerhouse, hydraulic steel structure, electromechanical equipment, switchyard, traffics, special equipment, safety monitoring system, working environment, safety management, and construction safety management of works and sections exceeding a certain scale with higher risk.

4.5.4 The hydropower project accidents shall be classified in accordance with the current national standard GB 6441, *The Classification for Casualty Accidents of Enterprise Staff and Workers*, and sorted according to risk level.

4.6 Recommendations on Safety Risk Countermeasures

4.6.1 The basis and principles for making suggestions on safety risk countermeasures, as well as the level order principle which safety technical measures shall follow shall be stated.

NOTE The level order principle which safety technical measures shall follow is elimination, prevention, mitigation, isolation, interlocking, and warning.

4.6.2 The safety pre-assessment report shall put forward technological and management countermeasures and suggestions for risk control based on hazardous and harmful factor identification results as well as qualitative and quantitative assessment results, following the principles of pertinence, technical feasibility, and economy.

4.6.3 The safety risk countermeasures may be categorized as "mandatory" or "optional" according to different risk levels.

4.6.4 Safety risk countermeasures shall be consistent with the analysis and assessment results of hazardous and harmful factors, and shall meet the following requirements:

1 With regard to the hazardous and harmful factors which are not under control, supplemental safety risk countermeasures shall be put forward in the pre-assessment report.

2 For the hazardous and harmful factors without safety control measures and major hazard installations, the requirements for adding safety facilities shall be put forward.

3 For the hazardous and harmful factors and major hazard installations for which safety management measures have been taken, the requirements for adding technical measures shall be put forward.

4 For the hazardous and harmful factors for which only indicative measures have been taken, indirect measures shall be added.

5 If conditions permit, the requirements for direct safety measures such as intrinsic safety, automatic control may be put forward.

6 Safety technical countermeasures shall be put forward in terms of general layout, function distribution, facilities, equipment, installations, etc. of the hydropower project.

7 Safety management countermeasures shall be put forward from the aspects of organizational setup, personnel management, emergency response plan management, occupational health management, etc. of the hydropower project.

8 Other safety protection measures shall be put forward from the needs of ensuring the safe operation of the hydropower project.

4.7 Principle and Framework for Preparation of Emergency Response Plan

4.7.1 The safety pre-assessment report of a hydropower project shall briefly describe the system structure and main content of emergency response plan for natural disasters and accidents.

4.7.2 Emergency response plans are classified into comprehensive emergency response plan, special emergency response plan, and on-site disposal plan. The main content of emergency response plan for natural disasters and accidents shall comply with the current national standard GB/T 29639, *Guidelines for Enterprises to Develop Emergency Response Plan for Work Place Accidents* and relevant sector regulations.

4.7.3 The safety pre-assessment report of a hydropower project shall explain the preparation, review, filing for record, drill, evaluation, revision, and other relevant requirements of emergency response plan.

4.7.4 The safety pre-assessment report of a hydropower project shall list the comprehensive emergency response plan, special emergency response plans, and on-site disposal plans that shall be prepared for the project.

4.8 Safety Cost Estimation

4.8.1 The safety pre-assessment report of a hydropower project shall briefly describe the preparation basis and price level year of the safety cost estimate.

4.8.2 The safety pre-assessment report of a hydropower project shall indicate the components of the safety cost. The safety cost shall include the following:

1 The funds for the implementation of safety measures and the improvement of work safety conditions.

2 The funds for the hazard control in workplace.

3 The funds for safety assessment, safety evaluation, work safety inspection, and testing.

4 The funds for the purchase, installation, and maintenance of the equipment and facilities for work safety and occupational hazard prevention and emergency rescue in workplaces.

5 The funds for work safety training and education, and work safety awards.

6 The funds for providing personnel protective equipment and occupational health examination.

7 The funds for establishing an emergency rescue system and carrying out emergency rescue drill.

8 The funds for paying work-related injury insurance premium for workers.

9 The fees for relevant emergency response plans and subject research.

10 The fees for special consultation, review, and acceptance of safety devices and measures.

4.8.3 The safety pre-assessment report of a hydropower project shall provide the bill of quantities for project safety. The bill of quantities include safety signs, safety protection facilities, working environment safety testing instruments, noise control, emergency equipment and facilities, relevant labor protection articles and personal protective equipment, safety tools, emergency supplies and equipment, safety education and training equipment, safety education classroom, sanitary room, etc. If any of the above items has been included in the relevant specialties, it need not be listed separately.

4.8.4 The safety pre-assessment report of a hydropower project shall present the list of safety cost estimates.

4.9 Assessment Conclusions

4.9.1 The safety pre-assessment report of a hydropower project shall present clear and definite conclusions.

4.9.2 The conclusions given in the safety pre-assessment report shall include the following:

1 List the main hazardous and harmful factors and give assessment results.

2 Identify the major risks of the hydropower project.

3 Define major hazard installations in the production process and their

risk levels.

4 Make suggestions on safety risk control measures to be highlighted.

5 Indicate whether the project complies with relevant national laws, regulations, and national or sector standards from the perspective of work safety.

6 Indicate whether the construction of the project is feasible in terms of safety.

4.10 Attachments and Attached Drawings

4.10.1 The attachments to the safety pre-assessment report of a hydropower project shall include the safety pre-assessment service agreement, the project development right granting document, review opinions on pre-feasibility study report, etc.

4.10.2 The attached drawings of the safety pre-assessment report shall include the project geographical location map, cascade development sketch, project layout plan, layouts and sections of main structures, main electrical connection diagram, main electromechanical equipment layout, switchyard layout, construction diversion layout, general construction layout, etc.

4.10.3 The design drawings attached to the safety pre-assessment report shall be signed as required.

Appendix A Data List for Safety Pre-assessment of Hydropower Project

1 General information

1.1 Profile of project owner

1.2 Project overview

1.3 Project general layout

1.4 Location relationship between project and surroundings

1.5 Geological and meteorological conditions

1.6 Staffing

2 Design basis and design documents

2.1 Engineering geological, hydrological and meteorological data on which project design is based

2.2 Other relevant safety information on which project design is based

2.3 Project development right granting document or project initiation approval document

2.4 Pre-feasibility study, feasibility study report interim results, and related drawings

2.5 Geological hazard assessment report and review opinions

2.6 Seismic safety assessment report and review opinions

2.7 Other design documents related to the project to be constructed or extended

3 Information on safety facilities and equipment

3.1 Description and explanation of production process

3.2 Description and explanation of safety facilities, equipment, and devices

4 Safety management organization and staffing

5 Safety cost

6 Related analogy information

6.1 Engineering analogy data

6.2 Related accident cases

7 Other data that can be used for safety pre-assessment

Appendix B Contents for Preparation of Safety Pre-assessment Report for Hydropower Project

1 Introduction

1.1 Purpose, scope and work process of pre-assessment

1.2 Assessment basis

1.3 Profile of project owner

2 Project overview

2.1 Brief introduction to project

2.2 Geographical location and surroundings

2.3 Hydrology and sediment

2.4 Engineering geology

2.5 Project tasks and scale

2.6 General layout of hydropower complex

2.7 Electromechanical equipment and hydraulic steel structures

2.8 Construction

2.9 Cost estimation

2.10 Project characteristics table

2.11 Relevant special study and main feasibility study results

3 Identification and analysis of hazardous and harmful factors

3.1 Basis for identification and analysis of hazardous and harmful factors

3.2 Hazardous and harmful factors in siting and general layout of hydropower complex

3.3 Hazardous and harmful factors in main buildings (structures)

3.4 Hazardous and harmful factors of main equipment

3.5 Hazardous and harmful factors in production process

3.6 Hazardous and harmful factors of natural disasters

3.7 Hazardous and harmful factors in workplaces

3.8 Hazardous and harmful factors during construction

3.9 Major hazard installations

4 Assessment Units and Assessment Methods

4.1 Definition of assessment units

4.2 Selection of assessment methods

4.3 Assessment methods used for each unit

5 Qualitative and quantitative assessment

5.1 Project siting and general layout

5.2 Near-dam reservoir bank

5.3 Water retaining structures

5.4 Navigation structures

5.5 Water release structures

5.6 Headrace system

5.7 Powerhouse and power generation system

5.8 Tailrace system

5.9 Switchyard

5.10 Access roads

5.11 Log pass

5.12 Fish pass

5.13 Special equipment

5.14 Safety monitoring system

5.15 Working environment

5.16 Construction

5.17 Safety management

6 Recommendations on safety risk countermeasures

6.1 Basis and principles for making suggestions on safety countermeasures

6.2 Safety technical countermeasures

6.3 Safety management countermeasures

6.4 Safety risk control measures for construction period

6.5 Other safety risk control measures

7 Principles and framework for preparation of emergency response plan

7.1 Composition and main content of emergency response plan

7.2 Emergency response plan preparation procedure

7.3 Emergency response plan required for the project

8 Safety cost estimation

8.1 Preparation basis

8.2 Price level year

8.3 Bill of quantities for safety

8.4 Safety cost estimate

Appendix C Format for Preparation of Safety Pre-assessment Report for Hydropower Project

C.0.1 The safety pre-assessment report of a hydropower project shall include the front cover, title page, preface, contents, main body, attachments, appendixes, etc.

C.0.2 The safety pre-assessment report of a hydropower project shall be in size A4, and bound on the left side.

C.0.3 The front cover of safety pre-assessment report of a hydropower projects shall include:

1 Name of entrusting party.

2 Name of project.

3 Title.

4 Name of safety assessment body.

5 Completion date of assessment report.

C.0.4 The front cover of safety pre-assessment report of a hydropower project shall be in accordance with Figure C.0.4.

Name of Entrusting Party

Name of Project

Safety Pre-assessment Report

Safety Assessment Body

Completion Date of Assessment Report

Figure C.0.4 Template of front cover

C.0.5 The title page of safety pre-assessment report of a hydropower project shall meet the following requirements:

1 The first title page shall carry the names of principals of safety assessment body such as legal person, technical director and assessment team leader, and the completion date and the official seal of the assessment body should be placed at the lower part of the page. See Figure C.0.5 for the template of first title page.

2 The second title page shall carry the names of assessment team members, technical experts, and other relevant persons. The assessment team members and technical experts shall sign their names by hand. The assessment team members shall be presented in accordance with Table C.0.5-1; the technical experts shall be presented in accordance with Table C.0.5-2.

Name of Entrusting Party
Name of Project

Safety Pre-assessment Report

Legal person:
Technical director:
Assessment team leader:

Completion date of assessment report
(Official seal of assessment body)

Figure C.0.5 Template of first title page

Table C.0.5–1 Assessment personnel

	Name	Qualification certificate No.	Work license No.	Signature
Team leader				
Team member				
Report drafter				
Report reviewer				
Principal for process control				
Technical director				

NOTE This form shall be filled out according to the actual number of personnel involved.

Table C.0.5–2 Technical experts

Name	Specialty	Professional title	Signature

NOTE List all experts when necessary.

Explanation of Wording in This Specification

1 Words used for different degrees of strictness are explained as follows in order to mark the differences in executing the requirements in this specification.

1) Words denoting a very strict or mandatory requirement:

"Must" is used for affirmation; "must not" for negation.

2) Words denoting a strict requirement under normal conditions:

"Shall" is used for affirmation; "shall not" for negation.

3) Words denoting a permission of a slight choice or an indication of the most suitable choice when conditions permit:

"Should" is used for affirmation, "should not" for negation.

4) "May" is used to express the option available, sometimes with the conditional permit.

2 "Shall meet the requirement of... " or "shall comply with... " is used in this specification to indicate that it is necessary to comply with the requirements stipulated in other relative standards and codes.

List of Quoted Standards

GB 6441, *The Classification for Casualty Accidents of Enterprise Staff and Workers*

GB/T 13861, *Classification and Code for the Hazardous and Harmful Factors in Process*

GB 18218, *Identification of Major Hazard Installations for Hazardous Chemicals*

GB/T 29639, *Guidelines for Enterprises to Develop Emergency Response Plan for Work Place Accidents*